Weird World of Animals

Glow-in-the-Dark Animals

By Mary Griffin

Gareth Stevens Publishing

Please visit our website, www.garethstevens.com. For a free color catalog of all our high-quality books, call toll free 1-800-542-2595 or fax 1-877-542-2596.

Names: Griffin, Mary, 1978-.
Title: Glow-in-the-dark animals / Mary Griffin.
Description: Buffalo, NY : Gareth Stevens Publishing, 2026. | Series: Weird world of animals | Includes glossary and index.
Identifiers: ISBN 9781482471830 (pbk.) | ISBN 9781482471847 (library bound) | ISBN 9781482471854 (ebook)
Subjects: LCSH: Bioluminescence--Juvenile literature.
Classification: LCC QH641.G75 2026 | DDC 572'.4358--dc23

Published in 2026 by
Gareth Stevens Publishing
2544 Clinton Street
Buffalo, NY 14224

Cover Designer: Nicholas Switalski
Interior Designer: Rachel Rising
Editor: Therese Shea

Photo credits: Cover, p. 1 Neil Bromhall/Shutterstock.com; Cover, pp. 1, 3-24 mycteria/Shutterstock.com; p. 4 klyaksun/Shutterstock.com; p. 5 Petar B photography/Shutterstock.com; p. 6 Freer/Shutterstock.com; p. 7 khlungcenter/Shutterstock.com; p. 8 IamTK/Shutterstock.com; p. 9 WUT.ANUNAI/Shutterstock.com; p. 11 IanRedding/Shutterstock.com; p. 13 Aleksei Miroliubov/Shutterstock.com; p. 15 Kan Sukarakan/ Shutterstock.com; p. 17 Khun Ta/Shutterstock.com; p. 19 Dmitriy Feldman svarshik/Shutterstock.com; p. 20 Kasefoto/Shutterstock.com; p. 21 Matis75/Shutterstock.com.

Printed in the United States of America

CPSIA compliance information: Batch #CSGS26: For further information contact Gareth Stevens, at 1-800-542-2595.

CONTENTS

*Words in the glossary appear in **bold** type the first time they are used in the text.*

AWESOME ADAPTATION

Do you think it would be handy to glow in the dark? You might not need a night-light anymore. It could make playing hide-and-seek a bit tricky, though.

Some animals can glow in the dark—really! An ability like this is called an adaptation. It's a feature that **evolved** in animals to help them live in the wild. In this book, you'll learn how glowing and other light-creating adaptations help certain species, or kinds, of animals survive.

Mushrooms aren't animals or plants—they're ***fungi***. However, some have an adaptation that helps them glow in the dark too.

That's WEIRD!

Scientists think glow-in-the-dark animals have existed for at least 540 million years!

Fireflies

A firefly might be the most common light-making bug near you. Fireflies don't glow in the dark, but they do produce flashes of light using a special **organ** under their **abdomen**. **Chemicals** in this organ can combine with the gas oxygen. Light is a result of the **reaction**.

A flashing light is a way that fireflies talk to each other. When a male firefly is looking for a **mate**, it flies near the ground and flashes its light about every six seconds.

Fireflies aren't flies—they're beetles!

That's WEIRD!

Firefly eggs and larvae glow too!

Female fireflies usually can't fly. On the ground, they flash their lights to the males when they're ready to mate. Every species of firefly has a special pattern of light flashing. There are 2,000 species in all! Light flashing is how species know each other. Firefly light can be blue-green, yellow, or orange.

Fireflies' lights may also be a warning to predators. Predators recognize that the flying beetles that make this light don't taste good or are even poisonous, so they stay away.

Animals that produce light in their bodies, like fireflies, are called bioluminescent (by-oh-loo-mih-NEH-sent).

That's WEIRD!

A group of fireflies may flash at the same time!

GLOWWORMS

Glowworms are crawling **insects** that give off light for long periods of time. Some are insect larvae. Others are wingless adult insects, such as female fireflies.

The glowworm that is sometimes called the railroad worm has 11 pairs of organs on its body. These organs glow yellow-green. Another pair of organs on its head glows red.

Scientists think glowworms' lights warn predators to stay away. They may be a way of saying, "I taste bad!" Or they may spook a predator.

This female common European glowworm glows about two hours each night until it attracts a mate.

That's WEIRD!

Some glowworms produce light to attract prey, or the creatures they eat.

COMB JELLIES

Comb jellies are invertebrates, or animals without a backbone, that look a bit like jelly ovals. Their name comes from the eight plates, called combs, that run along their body. They use these like oars to move through the water!

Most comb jellies glow blue or green. Some flash their light. Some release, or let go of, a glowing "ink" to defend themselves. Comb jellies use the same chemicals as fireflies do to make their light. However, calcium, not oxygen, causes the light-making reaction.

Comb jellies may glow to scare off predators.

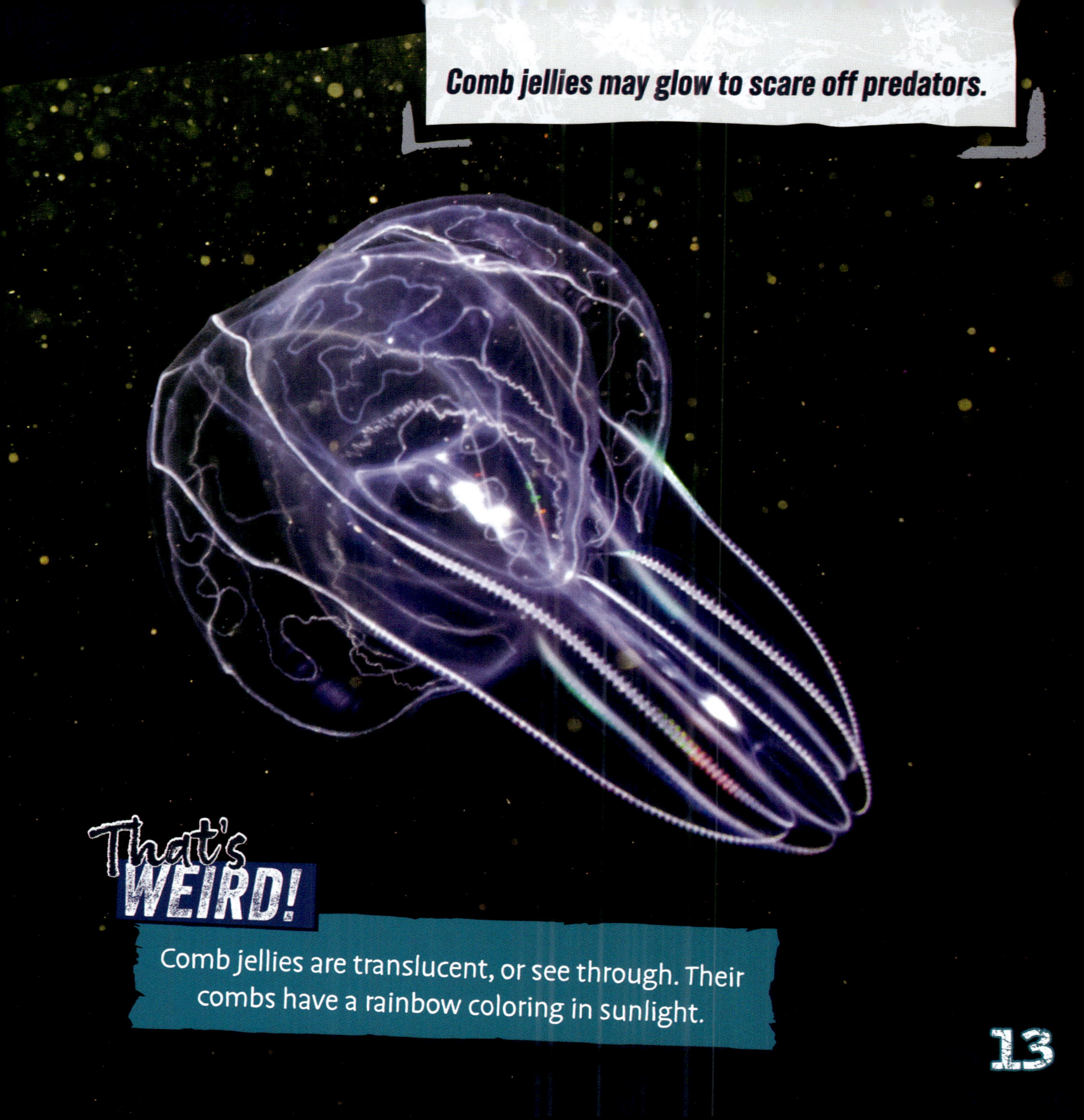

That's WEIRD!

Comb jellies are translucent, or see through. Their combs have a rainbow coloring in sunlight.

ANGLERFISH

There are 200 species of deep-sea anglerfish. Most live at the bottom of the Atlantic and Antarctic (or Southern) Oceans, where sunlight can't reach.

Female anglerfish have their own light. A piece of their back fin continues up over their head. At the end is a part called the esca, in which bacteria live. These bacteria glow, acting as a **lure** in the deep. The fish can make the lure move like a worm. When prey gets close to the lure, the fish sucks in the prey!

Anglerfish are some of the scariest looking fish in the deep.

That's WEIRD!

Male anglerfish grab on to female anglerfish. Their bodies fuse, or join, with the female's body, so they get food from her!

Firefly Squid

You have to travel down hundreds of feet in the western Pacific Ocean to find the firefly squid. Even then, you have to look hard because adults are only 3 inches (7.6 cm) long.

However, between March and June each year, thousands of firefly squid come up from the deep waters of the Sea of Japan to the surface. At night, they glow a beautiful bright blue to attract a mate. They use organs, called photophores, all over their body and **tentacles** to do this.

Scientists think firefly squids' glowing might also be a form of blending in to escape predators' notice. When looking at them from below, their shadow might be hard to see if they're glowing like the sunlight.

That's WEIRD!

Scientists think about three-fourths of all ocean creatures make some sort of light!

A DIFFERENT KIND OF LIGHT

Some kinds of animals aren't bioluminescent, which means they don't make their light in their bodies. Instead, they glow when viewed in a certain kind of light.

These animals' coverings—such as skin, feathers, or scales—absorb, or take in, **ultraviolet** light, which people can't see, and throw back a glowing light we can see. This adaptation, called biofluorescence (by-oh-flo-REH-sens), may help these animals attract prey or recognize each other. Scientists think some animals that are active at night see differently than we do. This adaptation may help them.

This* scorpion *appears to glow under ultraviolet light, showing the effects of biofluorescence.

GLOWING CATS?

Scientists have been studying the **genes** that make animals glow. They have been changing or adding genes to animals such as dogs, pigs, and monkeys that make them glow in ultraviolet light. Why? It helps them track what's going on in their bodies.

For example, scientists have been adding special cells to cats' bodies to help them fight off disease. The cells have the ability to glow. Scientists can track them and watch over the cats' health. Similar studies are helping people too!

Scientists are still finding animals that have a natural glow!

More Creatures That Glow

VAMPIRE SQUID lets off glowing snot to mix up predators

STOPLIGHT LOOSEJAW DRAGONFISH uses special red-light organs as a flashlight to spot prey

DUCK-BILLED PLATYPUS glows under ultraviolet light, possibly for predator protection

KITEFIN SHARK glows in the darkest ocean waters, possibly to spot prey

LANTERNFISH uses light to blend in with their surroundings and avoid predators

DUCK-BILLED PLATYPUS

GLOSSARY

abdomen: The final part of an insect's body that includes most of its guts and organs for mating.

chemical: Matter that can be mixed with other matter to cause changes.

evolve: To grow and change over time.

fungus: A living thing that is somewhat like a plant, but doesn't make its own food, have leaves, or have a green color. Fungi include molds and mushrooms.

gene: Tiny part of a cell that is passed along from one living thing to its offspring.

insect: A small, often winged, animal with six legs and three main body parts.

lure: An object used to bring an animal closer.

mate: One of two animals that come together to produce babies.

organ: A part inside an animal's body with a special function.

reaction: The way something responds.

scorpion: A small arachnid that lives in hot, dry areas and has a long body and a curved tail with a poisonous sting.

tentacle: A long, thin body part that sticks out from an animal's head or mouth.

ultraviolet: A range of wavelengths in light beyond the violet end of the visible color sequence.

FOR MORE INFORMATION

Books

Musgrave, Ruth. *Bioluminescent Animals*. New York, NY: DK Publishing, 2023.

Perdew, Laura. *Glowing Animals*. Parker, CO: The Child's World, 2025.

Websites

Bioluminescence
education.nationalgeographic.org/resource/bioluminescence/
Learn more about the science behind bioluminescence.

Top 10 Facts About Glow-in-the-Dark Creatures!
www.funkidslive.com/learn/top-10-facts/top-10-facts-about-glow-in-the-dark-creatures/
See how many of these facts you know!

INDEX